Edward R. Scheinerman

Approaching Asymptotics

Author

Edward R. Scheinerman
Department of Applied Mathematics and Statistics
Whiting School of Engineering
Johns Hopkins University
Baltimore, Maryland 21218 USA
www.ams.jhu.edu/ers

Cover

Jonah Scheinerman
www.scheinerman.net/jonah

ISBN: 978-1792819797

To Naomi and Yoni

Each piece, or part, of the whole of nature is always merely an approximation to the complete truth, or the complete truth so far as we know it. In fact, everything we know is only some kind of approximation because we know that we do not know all the laws as yet.

Richard Feynman

An approximate answer to the right question is worth far more than a precise answer to the wrong one.

John Tukey

The truth may be puzzling. It may take some work to grapple with. ...We have a method, and that method helps us to reach not absolute truth, only asymptotic approaches to the truth—never there, just closer and closer, always finding vast new oceans of undiscovered possibilities.

Carl Sagan

CONTENTS

1. Welcome

1.1. **When close enough is close enough.** An exact answer to a mathematical problem is usually the best answer but sometimes an exact answer is either unobtainable or unhelpful (or both).

Case in point: Suppose we are confronted with the equation $y = x^x$ and asked to solve for x. There's no nice way to do this algebraically. Of course, given a specific value for y (say, $y = 10$), we can use numerical methods to determine x (yielding $x = 2.50618...$). However, we'd like to know the general growth in x as y tends to infinity. See Exercise 42 on page 35.

At other times, we may have an exact answer, but it might contain too much irrelevant detail. For example, the n-th Fibonacci number is, exactly,

$$f_n = \frac{\left(\frac{1+\sqrt{5}}{2}\right)^n - \left(\frac{1-\sqrt{5}}{2}\right)^n}{\sqrt{5}}.$$

The formula is a bit complicated and that can mask its general behavior. Because $(1 - \sqrt{5})/2$ has absolute value less than 1, its contribution is tiny when n is large. Taking that into account, the formula becomes much simpler as we condense it like this:

$$f_n \sim \frac{\varphi^n}{\sqrt{5}}$$

where φ is the golden mean, $(1 + \sqrt{5})/2 \approx 1.618$.

Asymptotic notation and methods are fabulous tools for providing useful approximations in situations where exact results are unobtainable and/or too cluttered with unimportant detail.[1] The use of asymptotics arises in many branches of mathematics and computer science. A prime example from number theory concerns the distribution of primes. The function $\pi(n)$ is defined to be the number of prime numbers between 1 and n. With a modest amount of computation, we can determine that the number of primes up to one billion is

$$\pi(1{,}000{,}000{,}000) = 50{,}847{,}534.$$

How many primes have, say, at most 100 digits? Determining the exact value of $\pi(10^{100})$ is computationally infeasible, but the

---

[1]This is reminiscent of stores charging $19.99 for an item. It'd be simpler (and less irritating) to charge $20.

celebrated Prime Number Theorem gives us an excellent approximation:

$$\pi(n) \sim \frac{n}{\log n}.$$

When n is one billion, this approximation gives 48 million which is about 5% off from the true value (slightly more than 50 million). For the case when n is a googol, we find that $\pi(n) \approx 4.34 \times 10^{97}$.

Our goal is to provide a quick introduction to asymptotics. More advanced methods can be found in other books such as [1] and the lovely *Asymptopia* [4]. Asymptotic notation and a gentler introduction can also be found in [2] and [3].

We certainly hope that you find the presentation of this material is clear and helpful. However, the best way to learn is to do, and for this we provide a wide variety of exercises with complete solutions.

1.2. **Thanks.** I received wonderful feedback and suggestions from students and colleagues in the preparation of this book. Both Jim Fill and Daniel Scheinerman read early versions deeply and gave extensive comments; their input touched every corner of this work.

Thanks to Jonah Scheinerman for the wonderful cover art.

I also want to thank the following people who also gave a variety of suggestions and uncovered errors: Tara Abrishami, Daniel Cranston, Jonathan Katzman, Zeinab Maleki, Elizabeth Reiland, Kate Richer, Jonah Scheinerman, and Yiguang Zhang.

2. Notation

Consider the simple binomial coefficient $\binom{n}{2}$. We know that it equals $\frac{1}{2}n(n-1)$. However, when n is large, it is accurately approximated by $n^2/2$. When $n = 1000$, the relative error between the true and the approximate values is about 0.1%. The goal of this section is to develop notation that makes "approximately" precise.

2.1. Asymptotic equality. The general setup is that we have two functions f and g with either

$$f, g : \mathbb{Z}^+ \to \mathbb{R} \qquad \text{or} \qquad f, g : \mathbb{R} \to \mathbb{R}.$$

This assumption is tacit in all the definitions that follow in this section.

We do not use the symbol $\approx$ except in a vague, undefined sense of "is approximately equal to". (We have much better notation coming.) When we say $f(x) \approx g(x)$ we need to set one of two contexts: x small or x large. Generally this means that some condition holds as $x \to 0$ or $x \to \infty$, respectively[2].

Definition 1 (Asymptotically equal). Let f and g be functions. We say that f is *asymptotically equal* to g, and we write $f \sim g$, provided

for x large: $\displaystyle\lim_{x\to\infty} \frac{f(x)}{g(x)} = 1$ or for x small: $\displaystyle\lim_{x\to 0} \frac{f(x)}{g(x)} = 1.$

For example $\binom{n}{2} \sim \frac{1}{2}n^2$. Implicit in this claim is that the expressions on the two sides of the $\sim$ symbol are *functions* of n (and not specific numbers). Also implicit is that we are asserting this relation for n large. And here's the justification:

$$\lim_{n\to\infty} \frac{\binom{n}{2}}{\frac{1}{2}n^2} = \lim_{n\to\infty} \frac{n(n-1)/2}{n^2/2} = 1.$$

Asymptotic formulas may involve more than one variable name. In such instances, it is important to understand which letter stands for the name of a number and which is a function. For example, consider this statement:

---

[2]In the case of "x small", the functions are defined for real inputs. For "x large", we may have either $f : \mathbb{R} \to \mathbb{R}$ or $f : \mathbb{Z}^+ \to \mathbb{R}$.

Statement. *For k fixed and n large, $\binom{n}{k} \sim n^k/k!$.*

Here is a precise unraveling of the statement's meaning:

Translation. *Let k be a nonnegative integer. The functions $f(n) = \binom{n}{k}$ and $g(n) = n^k/k!$ are asymptotically equal for n large. That is:*

$$\forall k \in \mathbb{N}, \ \lim_{n \to \infty} \frac{\binom{n}{k}}{n^k/k!} = 1.$$

By saying "for k fixed" we are indicating that k is a specific integer that does not vary with n.

2.2. **Big oh.** The big oh notation, $O(\cdot)$, is used to indicate a missing term in an expression and to provide a bound on how large that term is. For example, let's revisit the binomial coefficient $\binom{n}{2}$. We know that $\binom{n}{2} \sim \frac{1}{2}n^2$. We could be completely precise by writing

$$\binom{n}{2} = \frac{1}{2}n^2 - \frac{1}{2}n.$$

This reveals the exact difference between $\binom{n}{2}$ and $\frac{1}{2}n^2$. However, in many instances it is not necessary (nor desirable) to be so precise. We simply want to give a bound on how large the difference is between the two terms. We express this as follows:

$$\binom{n}{2} = \frac{1}{2}n^2 + O(n)$$

where $O(n)$ indicates a missing term[3] that grows at most linearly in n.

Definition 2 (Big oh, large argument). Let f and g be functions from $\mathbb{Z}^+$ to $\mathbb{R}$. We say that f is $O(g)$ provided there is

---

[3]We may also write $\binom{n}{2} = O(n^2)$; this means that the function $f(n) = \binom{n}{2}$ grows at most quadratically with n. In this case, we may replace "$=$" with "is" and write $\binom{n}{2}$ is $O(n^2)$. Of course, this is less precise than writing $\binom{n}{2} = \frac{1}{2}n^2 + O(n)$.

a positive number C such that for all n sufficiently large we have

$$|f(n)| \leq C|g(n)|.$$

The phrase "for all n sufficiently large" may also be expressed as "with at most finitely many exceptions." Or, with complete precision, this can be written:

$$\exists C > 0, \ \exists n_0 \in \mathbb{Z}^+, \ \forall n \geq n_0, \ |f(n)| \leq C|g(n)|.$$

Definition 2 only considers functions defined on the positive integers. However, we may certainly use the big-oh notation for functions defined on all real numbers. In which case, $f(x) = O(g(x))$ translates to this:

$$\exists C > 0, \ \exists x_0 \in \mathbb{R}, \ \forall x \geq x_0, \ |f(x)| \leq C|g(x)|.$$

Just as asymptotic equality has two version for large and small arguments, so too we can use the big-oh notation for small arguments.

For example, when x is near zero, e^x is approximately $1 + x$. More precisely, we may write $e^x = 1 + x + O(x^2)$. The $O(x^2)$ stands for a terms that is bounded by a constant times x^2. Here is the precise meaning.

Definition 3 (Big oh, small argument). Let $f, g : \mathbb{R} \to \mathbb{R}$ be functions. We say that $f(x)$ is $O(g(x))$ provided there are positive numbers C and ε such that for all x with $-\varepsilon < x < \varepsilon$ we have

$$|f(x)| \leq C|g(x)|.$$

Thus, when we say that $e^x = 1 + x + O(x^2)$ we mean that there is a positive number C such that for all x near 0, we have

$$|e^x - x - 1| \leq Cx^2.$$

This is illustrated in Figure 1 on the next page.

A close relative to the big-oh notation is the big-omega notation. When we say that f is $\Omega(g)$, this means that g is $O(f)$. In other words, for large n there is a positive number C and an integer

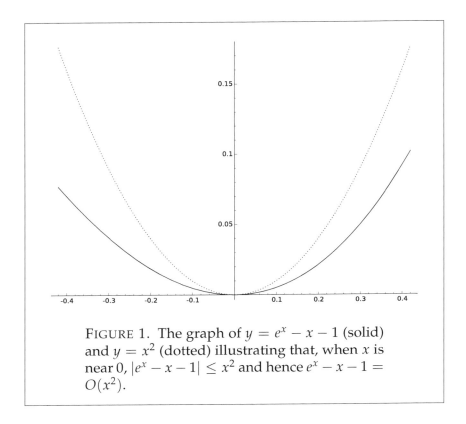

FIGURE 1. The graph of $y = e^x - x - 1$ (solid) and $y = x^2$ (dotted) illustrating that, when x is near 0, $|e^x - x - 1| \leq x^2$ and hence $e^x - x - 1 = O(x^2)$.

n_0 such that for all $n \geq n_0$ we have $|f(n)| \geq C|g(n)|$. And for small x, it means that there are positive numbers C and ε such that $|f(x)| \geq C|g(x)|$ for all $-\varepsilon < x < \varepsilon$.

In summary, the big-oh notation is an implicit upper bound and the big-omega notation is a lower bound. Imprecisely, we have

$$f = O(g) \implies |f(x)| \leq C|g(x)| \qquad \text{and}$$
$$f = \Omega(g) \implies |f(x)| \geq C|g(x)|$$

for all "appropriate" values of x (large or small) and where $C > 0$.

2.3. **Big theta.** The big-theta notation, $\Theta(\cdot)$, is a close relative to the big-oh notation. Let f and g be functions. When we say that f is $\Theta(g)$ we mean that both f is $O(g)$ and g is $O(f)$. Equivalently, when f is $\Theta(g)$ we mean that there are positive constants A and B

such that for all n sufficiently large

$$A \le \left| \frac{f(n)}{g(n)} \right| \le B.$$

Thus big-oh notation is an upper bound while big theta implies upper and lower bounds. Some examples:

- $\binom{n}{2} = O(n^2)$ is true.
- $\binom{n}{2} = O(n^3)$ is also true.
- But $\binom{n}{2} = O(n)$ is false.
- $\binom{n}{2} = \Theta(n^2)$ is true.
- But $\binom{n}{2} = \Theta(n^k)$ is false for any integer $k \ne 2$.

There is also a small-argument version of big theta. In this context, $f = \Theta(g)$ means there are positive constants A and B such that for all x sufficiently near 0, we have

$$A \le \left| \frac{f(x)}{g(x)} \right| \le B.$$

Notice that $f = \Theta(g)$ if and only if $g = \Theta(f)$.

There is an alternative notation that carries the same meaning as big theta. When we write $f \asymp g$ this means $f = \Theta(g)$. Notice that $\asymp$ is an equivalence relation.

One final observation: The $\asymp$ relation is a relaxation of the $\sim$ relation: $f \sim g \Rightarrow f \asymp g$, but the converse does not hold. See Exercise 4 on page 31.

2.4. Little oh.

The little-oh notation is used to indicate that one function is "much smaller" than another.

Definition 4 (Little oh, large argument). Let $f, g : \mathbb{Z}^+ \to \mathbb{R}$ be functions. We say that f is $o(g)$ provided

$$\lim_{n \to \infty} \frac{f(n)}{g(n)} = 0.$$

For example, $\binom{n}{2} = \frac{1}{2}n^2 + o(n^2)$.

Definition 4 is expressed in terms of functions defined on the positive integers. It is also reasonable to apply this notation for functions defined for real values $f, g : \mathbb{R} \to \mathbb{R}$.

There is also a small-argument version of the little-oh notation.

Definition 5 (Little oh, small argument). Let $f, g : \mathbb{R} \to \mathbb{R}$ be functions. We say that f is $o(g)$ provided

$$\lim_{x \to 0} \frac{f(x)}{g(x)} = 0.$$

The expression $o(1)$ appears often in formulas and signifies a "negligible term." Of course, the precise meaning is that the missing term tends to zero as $n \to \infty$ or $x \to 0$. For example,

$$\sqrt{n-1} = \sqrt{n} + o(1). \tag{1}$$

Of course $\sqrt{n-1} < \sqrt{n}$, so the missing term is negative. Writing "$+o(1)$" is proper, but it may be more comfortable to write this as $\sqrt{n-1} = \sqrt{n} - o(1)$. See Exercise 36 on page 34.

There is a reciprocal[4] notation to little oh: little omega.

Definition 6 (Little omega). Let f, g be real-valued functions. We say that f is $\omega(g)$ exactly when f/g tends to $+\infty$.

In other words, when we say that f is $\omega(g)$ we mean:

$$\lim_{n \to \infty} \frac{f(n)}{g(n)} = +\infty \ \text{(large argument)} \quad \text{or}$$

$$\lim_{x \to 0} \frac{f(x)}{g(x)} = +\infty \ \text{(small argument)}.$$

The expression $\omega(1)$ indicates any function that tends to positive infinity.

In conclusion, when $f = o(g)$, we are saying that f is much smaller than g, and when $f = \omega(g)$, we are saying that f is much larger than g. There's handy notation for this. Assuming f and g only take on positive values:

$$f \ll g \iff f = o(g) \iff g = \omega(f)$$
$$\text{and} \quad f \gg g \iff f = \omega(g) \iff g = o(f).$$

---

[4]The relation is only *mostly* reciprocal. See Exercise 29 on page 33.

3. Methods

It's not difficult to show that $\sim$, is-asymptotically-equal-to, is an equivalence relation (Exercise 30 on page 33). This does not mean that we may freely substitute asymptotically equal expressions for each other and expect the results to be asymptotically equal (see, for example, Exercises 34 and 35 on page 34).

3.1. Asymptotics and basic operations. It's elementary that if $a = x$ and $b = y$, then $a + b = x + y$, $ab = xy$, and so forth. What happens when we replace $=$ with $\sim$?

> **Proposition 7.** Let f_1, f_2, g_1, g_2 be real-valued functions with $f_1 \sim g_1$ and $f_2 \sim g_2$. Then $f_1 f_2 \sim g_1 g_2$.

Proof. Given $f_1 \sim g_1$ and $f_2 \sim g_2$, we have

$$\lim \left[\frac{f_1(x)f_2(x)}{g_1(x)g_2(x)} \right] = \lim \left[\frac{f_1(x)}{g_1(x)} \right] \cdot \lim \left[\frac{f_2(x)}{g_2(x)} \right] = 1 \cdot 1 = 1$$

and so $f_1 f_2 \sim g_1 g_2$. $\qquad\square$

> **Proposition 8.** Let f and g be real-valued functions with $f \sim g$. Then $1/f \sim 1/g$.

Proof. If $f \sim g$, then $\lim \frac{f(x)}{g(x)} = 1$, which implies that $\lim \frac{g(x)}{f(x)} = 1$. But then

$$\lim \frac{1/f(x)}{1/g(x)} = \lim \frac{g(x)}{f(x)} = 1$$

and so $1/f \sim 1/g$. $\qquad\square$

Taken together, Propositions 7 and 8 give the following: If $f_1 \sim g_1$ and $f_2 \sim g_2$, then $\frac{f_1}{f_2} \sim \frac{g_1}{g_2}$.

> **Proposition 9.** Let f_1, f_2, g_1, g_2 be positive real-valued functions with $f_1 \sim g_1$ and $f_2 \sim g_2$. Then $f_1 + f_2 \sim g_1 + g_2$.

15

Note that we require the functions to be positive; see Exercise 32 on page 33.

The proof rests on this simple idea. We are given that $f_1/g_1 \to 1$ and $f_2/g_2 \to 1$. We prove that $(f_1 + f_2)/(g_1 + g_2) \to 1$ by showing that $(f_1 + f_2)/(g_1 + g_2)$ lies between f_1/g_1 and f_2/g_2.

Proof. For any given value of x we have

$$\frac{f_1(x)}{g_1(x)} \leq \frac{f_2(x)}{g_2(x)} \qquad \text{or} \qquad \frac{f_1(x)}{g_1(x)} \geq \frac{f_2(x)}{g_2(x)}.$$

The first inequality implies $f_1(x)g_2(x) \leq f_2(x)g_1(x)$. Adding $f_1(x)g_1(x)$ to both sides gives $f_1(x)g_1(x) + f_1(x)g_2(x) \leq f_1(x)g_1(x) + f_2(x)g_1(x)$. Divide by $g_1(x)[g_1(x) + g_2(x)]$ to get

$$\frac{f_1(x)g_1(x) + f_1(x)g_2(x)}{g_1(x)[g_1(x) + g_2(x)]} \leq \frac{f_1(x)g_1(x) + f_2(x)g_1(x)}{g_1(x)[g_1(x) + g_2(x)]}$$

which simplifies to

$$\frac{f_1(x)}{g_1(x)} \leq \frac{f_1(x) + f_2(x)}{g_1(x) + g_2(x)}.$$

Likewise, adding $f_2(x)g_2(x)$ to both sides of $f_1(x)g_2(x) \leq f_2(x)g_1(x)$ and then dividing by $g_2(x)[g_1(x) + g_2(x)]$ yields

$$\frac{f_1(x) + f_2(x)}{g_1(x) + g_2(x)} \leq \frac{f_2(x)}{g_2(x)}.$$

In summary:

$$\frac{f_1(x)}{g_1(x)} \leq \frac{f_2(x)}{g_2(x)} \implies \frac{f_1(x)}{g_1(x)} \leq \frac{f_1(x) + f_2(x)}{g_1(x) + g_2(x)} \leq \frac{f_2(x)}{g_2(x)}.$$

By an analogous argument, we have

$$\frac{f_1(x)}{g_1(x)} \geq \frac{f_2(x)}{g_2(x)} \implies \frac{f_1(x)}{g_1(x)} \geq \frac{f_1(x) + f_2(x)}{g_1(x) + g_2(x)} \geq \frac{f_2(x)}{g_2(x)}.$$

In both cases, the ratio $[f_1(x) + f_2(x)]/[g_1(x) + g_2(x)]$ lies in the interval between $f_1(x)/g_1(x)$ and $f_2(x)/g_2(x)$. Since these latter two fractions tend to 1, we have

$$\lim \frac{f_1(x) + f_2(x)}{g_1(x) + g_2(x)} = 1$$

and conclude $f_1 + f_2 \sim g_1 + g_2$. $\qquad \square$

The results of Propositions 7–9 also hold for the $\asymp$ relation. The proofs are nearly identical.

Proposition 10. *Let f_1, f_2, g_1, g_1 be real-valued functions with $f_1 \asymp g_1$ and $f_2 \asymp g_2$. Then*

- $f_1 f_2 \asymp g_1 g_2$,
- $1/f_1 \asymp 1/g_1$, and
- *if the functions are positive, $f_1 + f_2 \asymp g_1 + g_2$.* □

Not surprisingly, we have similar results for $\ll$.

Proposition 11. *Let f_1, f_2, g_1, g_1 be positive real-valued functions with $f_1 \ll g_1$ and $f_2 \ll g_2$. Then*

- $f_1 f_2 \ll g_1 g_2$,
- $1/f_1 \gg 1/g_1$, and
- $f_1 + f_2 \ll g_1 + g_2$. □

Sometimes one may infer from $f \sim g$ that $\log f \sim \log g$. Of course, for this to make sense we need that f and g are positive. Because log is continuous we have

$$f \sim g \quad \implies \quad \lim_x \frac{f(x)}{g(x)} = 1 \quad \implies \quad \lim_x \log\left[\frac{f(x)}{g(x)}\right] = 0$$

$$\implies \quad \lim_x \left[\log f(x) - \log g(x)\right] = 0.$$

Now suppose that $\log g$ is bounded away from 0. Then

$$0 = \lim_x \left[\frac{\log f(x) - \log g(x)}{\log g(x)}\right] = \lim_x \left[\frac{\log f(x)}{\log g(x)} - 1\right]$$

$$\implies \quad \lim_x \frac{\log f(x)}{\log g(x)} = 1$$

and hence $\log f \sim \log g$. The same conclusion holds if $\log f(x)$ is bounded away from 0.

On the other hand, suppose $f(x)$ and $g(x)$ are differentiable functions that both approach 0 or both tend to ∞. By l'Hôpital's

Rule[5] we have

$$\lim_x \frac{\log f(x)}{\log g(x)} = \lim_x \frac{f'(x)/f(x)}{g'(x)/g(x)} = \lim_x \frac{f'(x)}{g'(x)} \cdot \lim_x \frac{g(x)}{f(x)} = 1 \cdot 1 = 1$$

and we conclude $\log f(x) \sim \log g(x)$.

The general implication $f \sim g \Rightarrow \log f \sim \log g$ is false (see Exercise 34 on page 34) but often holds. Usually, but not always, taking logarithms makes expressions "more" asymptotic, not less. For example, let $f(n) = n^2 \log n$ and let $g(n) = n^2$. Clearly $f(n) \gg g(n)$. Observe that

$$\log f(n) = 2 \log n + \log \log n \qquad \text{and} \qquad \log g(n) = 2 \log n$$

and

$$\lim_{n \to \infty} \frac{2 \log n + \log \log n}{2 \log n} = \lim_{n \to \infty} \left[1 + \frac{\log \log n}{2 \log n} \right] = 1.$$

Therefore $\log f \sim \log g$.

The reverse implication, $\log f \sim \log g \Rightarrow f \sim g$ does not hold; see Exercise 35 on page 34.

3.2. **Factors of one.** Let's consider a special case of Proposition 11:

Proposition 12. *Let f, g, and h be functions such that $h(x) = f(x)g(x)$ and $g(x) \sim 1$. Then $f(x) \sim h(x)$.* $\square$

The statement $g(x) \sim 1$ means that $\lim g(x) = 1$ (where $x \to 0$ or $x \to \infty$ depending on the context). In other words, if we multiply a function by a factor that tends to 1, then, asymptotically, we have not changed the expression. For example, consider this expression:

$$f(n) = \left(\frac{n}{n} \right) \left(\frac{n-1}{n} \right) \left(\frac{n-2}{n} \right) \left(\frac{n-3}{n} \right) \left(\frac{n-4}{n} \right). \tag{2}$$

Each factor in (2) is asymptotically equal to one (for n large) and therefore the overall product is asymptotically equal to one: $f(n) \sim 1$.

---

[5]Note that we used l'Hôpital's Rule twice. The first application is to the fraction $\log f(x) / \log g(x)$. The second application is to conclude that $f'(x)/g'(x) \to 1$ because we are given that $f \sim g$.

However, consider this expression:

$$g(n) = \left(\frac{n}{n}\right)\left(\frac{n-1}{n}\right)\left(\frac{n-2}{n}\right)\cdots\left(\frac{n-\lceil n/\log n\rceil}{n}\right). \quad (3)$$

Although every factor in (3) is asymptotically equal to 1, the overall function $g(n)$ is not! Why? In (2), the function f is the product of a specific number of factors *that does not depend on n*. Thus we may apply Proposition 12 a specific number of times. However, in (3) the number of factors is unbounded and depends on n. See Exercise 33 on page 34.

4. FORMULAS

4.1. Taylor series. Showing asymptotic equality of two functions boils down to evaluating a limit. Taylor series provide a handy way to derive asymptotic formulas. Let $f : \mathbb{R} \to \mathbb{R}$ be a function. Assuming it has derivatives of all orders at 0, its Taylor series is

$$f(0) + f'(0)x + \frac{1}{2!}f''(0)x^2 + \frac{1}{3!}f'''(0)x^3 + \cdots .$$

Often (but not always) this power series converges to $f(x)$ when x is near 0. In this case f is called *analytic* at 0. If we truncate the Taylor series at, say, the x^k term we have an approximation to $f(x)$ for x near 0. The difference between the true value, $f(x)$, and its approximation is $O(x^{k+1})$.

Theorem 13. *Let $f : \mathbb{R} \to \mathbb{R}$ be analytic at 0. For every positive integer k we have*

$$f(x) = \sum_{j=0}^{k} \frac{f^{(j)}(0)}{j!}x^j + O\left(x^{k+1}\right). \qquad \square$$

Note: In case f is analytic at some other value, say at a, we may write

$$f(x) = \sum_{j=0}^{k} \frac{f^{(j)}(a)}{j!}(x-a)^j + O\left[(x-a)^{k+1}\right].$$

Here, the big-oh term is understood as applying for x near a. That is, there is a constant C such that for all x sufficiently near a, the error term is bounded by $C\left|(x-a)^{k+1}\right|$.

Three functions deserve special attention: $\exp(x)$ (i.e., e^x), $\log x$ (base-e logarithm), and $1/(1-x)$ (the sum of a geometric series). Here are their Taylor series.

$$\exp(x) = e^x = 1 + x + \frac{x^2}{2!} + \frac{x^3}{3!} + \cdots \tag{4}$$

$$\log(1 - x) = -x - \frac{x^2}{2} - \frac{x^3}{3} - \frac{x^4}{4} - \cdots \tag{5}$$

$$\frac{1}{1 - x} = 1 + x + x^2 + x^3 + \cdots \tag{6}$$

For example, if we apply Theorem 13 to $\exp(x)$ we have:

$$\exp(x) = 1 + x + O(x^2).$$

There's an inequality version of this approximation for exp as well:

Proposition 14. *For all $x \in \mathbb{R}$, $e^x \geq 1 + x$.*

Figure 2 on the next page illustrates this inequality graphically. We give an analytic proof.

Proof. Let $f(x) = e^x - 1 - x$ and we work to show that $f(x) \geq 0$ for all x.

Consider $f'(x) = e^x - 1$. Notice that

$$f'(x) \begin{cases} < 0 & \text{for } x < 0, \\ = 0 & \text{for } x = 0, \text{ and} \\ > 0 & \text{for } x > 0. \end{cases}$$

Thus $f(x)$ decreases for x negative, is 0 at $x = 0$, and increases for $x > 0$, and this implies $f(x) \geq 0$ for all x. $\qquad\square$

Equation (6) can be seen as a special case of the Generalized Binomial Theorem (presented on the following page).

One way to express the standard Binomial Theorem is that for $n \in \mathbb{N}$

$$(1 + x)^n = \sum_{k=0}^{n} \binom{n}{k} x^k.$$

21

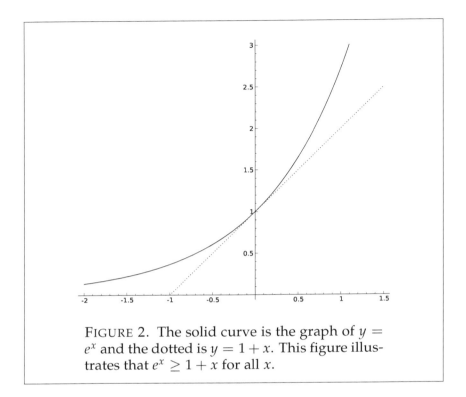

FIGURE 2. The solid curve is the graph of $y = e^x$ and the dotted is $y = 1 + x$. This figure illustrates that $e^x \geq 1 + x$ for all x.

The generalized version makes two modifications. First, the upper index of the sum may be changed from n to ∞ since the binomial coefficient $\binom{n}{k}$ is equal to zero when $k > n$. Second, and more significantly, we allow n to be any real number. That requires us to understand the binomial coefficient more broadly. Write the binomial coefficient like this:

$$\binom{n}{k} = \frac{1}{k!}n(n-1)(n-2)\cdots(n-k+1).$$

Thinking of k as a constant, this is a degree-k polynomial in n. Hence, there is no problem replacing n with an arbitrary real number α:

$$\binom{\alpha}{k} = \frac{1}{k!}\alpha(\alpha-1)(\alpha-2)\cdots(\alpha-k+1).$$

Theorem 15 (Generalized Binomial Theorem). *Let $\alpha, x \in \mathbb{R}$. Then*

$$(1+x)^\alpha = \sum_{k=0}^{\infty} \binom{\alpha}{k} x^k. \qquad \square$$

The sum is finite if $\alpha \in \mathbb{N}$. Otherwise the series converges when $|x| < 1$.

If we truncate the sum, we have the following asymptotic result.

Proposition 16. *Let $\alpha \in \mathbb{R}$ and $n \in \mathbb{N}$. Then for x small we have*

$$(1+x)^\alpha = \sum_{k=0}^{n} \binom{\alpha}{k} x^k + O(x^{n+1}). \qquad \square$$

4.2. Exponentiation and logarithm.

The exp function may be defined in a variety of ways including these:

$$\exp(x) = e^x = \lim_{n \to \infty} \left(1 + \frac{x}{n}\right)^n = \sum_{k=0}^{\infty} \frac{x^k}{k!}.$$

Implicit in these expressions is that x is a given real (or complex) number. We may be tempted to replace x in the limit expression with a function and hope that $\left[1 + \frac{f(n)}{n}\right]^n$ is asymptotic to $\exp[f(n)]$. The following result shows when we may draw that conclusion.

Proposition 17. *Let $f : \mathbb{N} \to \mathbb{R}$ and suppose that $f(n) = o(\sqrt{n})$. Then, for n large, we have*

$$\left(1 + \frac{f(n)}{n}\right)^n \sim \exp[f(n)].$$

Proof. Define

$$F(n) = \left(1 + \frac{f(n)}{n}\right)^n$$

23

and take logs to get

$$\log F(n) = n \log \left(1 + \frac{f(n)}{n} \right).$$

Using the fact that $\log(1 + y) = y - O(y^2)$ we have

$$\log F(n) = n \cdot \frac{f(n)}{n} - n \cdot O \left(\frac{[f(n)]^2}{n^2} \right)$$

$$= f(n) - O \left(\frac{[f(n)]^2}{n} \right) = f(n) + o(1).$$

Exponentiating gives

$$F(n) = \exp \left[f(n) + o(1) \right] = e^{f(n)} \cdot e^{o(1)} \sim e^{f(n)}. \qquad \square$$

Proposition 17 does not hold if $f(n)$ is too large; see Exercise 28 on page 33. However, we do have the following inequality.

Proposition 18. *Let $n \in \mathbb{Z}^+$ and $x \in \mathbb{R}$ with $1 + \frac{x}{n} > 0$. Then $\left(1 + \frac{x}{n} \right)^n \leq e^x$.*

Proof. Exercise 26 on page 33. $\qquad \square$

4.3. **Factorial.** Recall that for $n \in \mathbb{N}$ we have

$$n! = \begin{cases} 1 & \text{for } n = 0 \text{ and} \\ n \cdot (n-1)! & \text{for } n > 0. \end{cases}$$

That is, $n! = n \cdot (n-1) \cdots 3 \cdot 2 \cdot 1$. The factorial function appears often in mathematics and it is extremely useful to have an accurate approximation.

Theorem 19 (Stirling's Formula). *For n large, $n! \sim \sqrt{2\pi n}(n/e)^n$.* $\qquad \square$

We don't include a proof of this formula, but we can give a simple derivation that gets us close. We want an approximation for

$$n! = 1 \times 2 \times 3 \times \cdots \times n.$$

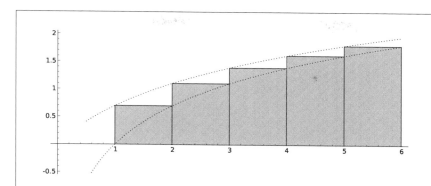

FIGURE 3. Approximating $\log n!$ from above and below by integrals. The shaded rectangles have total area $\log 1 + \log 2 + \cdots + \log 6 = \log 6!$. The lower dotted curve is $y = \log x$ and the upper dotted curve is $y = \log(x+1)$.

Taking logs gives

$$\log n! = \log 1 + \log 2 + \log 3 + \cdots + \log n.$$

This sum can be approximated by integrals. Take a close look at Figure 3 to see that

$$\int_1^n \log x\, dx < \log n! < \int_1^n \log(x+1)\, dx.$$

Recall that $\int \log x\, dx = x\log(x) - x$ and we have

$$n\log(n) - n + 1 < \log n! < (n+1)\log(n+1) - n - 2\log(2) + 1.$$

Now, let's take the average of the lower and upper bounds in this last equation to get the following approximation:

$$\log n! \approx \frac{n\log n + (n+1)\log(n+1)}{2} - n - \log 2 + 1$$

$$= \frac{n}{2}\log\left[n^2(1+\tfrac{1}{n})\right] + \frac{1}{2}\log(n+1) - n - \log 2 + 1$$

$$= n\log n + \frac{n}{2}\log(1+\tfrac{1}{n}) + \frac{1}{2}\log(n+1) - n - \log 2 + 1.$$

We now exponentiate both sides. Here's what happens to the terms:

$$n \log n \longrightarrow n^n$$

$$\frac{n}{2} \log(1 + \tfrac{1}{n}) \longrightarrow \left(1 + \frac{1}{n}\right)^{n/2} \sim \sqrt{e}$$

$$\frac{1}{2} \log(n+1) \longrightarrow \sqrt{n+1} \sim \sqrt{n}$$

$$-n \longrightarrow e^{-n}$$

$$-\log 2 + 1 \longrightarrow e/2.$$

Multiplying these together gives the following approximation:

$$n! \approx C\sqrt{n} \left(\frac{n}{e}\right)^n$$

where $C = e^{3/2}/2 \approx 2.24$. This formula is nearly correct; only the constant is wrong. It should be $\sqrt{2\pi} \approx 2.51$.

4.4. Falling factorial.

In combinatorial mathematics we often see expressions of the form $n(n-1)(n-2)\ldots(n-k+1)$ that we can think of as "partial" factorials: the factors don't go all the way down to 1. We call expressions such as these *falling factorial* which we define as follows.

Definition 20. Let $x \in \mathbb{R}$ and $k \in \mathbb{N}$. The *falling factorial* $x^{\underline{k}}$ is defined as

$$x^{\underline{k}} = \underbrace{x(x-1)(x-2)\cdots(x-k+1)}_{k \text{ terms}}.$$

For $n \in \mathbb{N}$, the expression $n^{\underline{k}}$ gives the number of repetition-free length-k sequences consisting of elements drawn from an n-element set. Another popular notation for the falling factorial is $(n)_k$.

We make note of a few special cases.

- $x^{\underline{0}} = 1$ because, by definition, empty products are set equal to 1.
- For a nonnegative integer n, we have $n^{\underline{n}} = n!$.
- If n is a nonnegative integer and $k > n$, then $n^{\underline{k}} = 0$.

26

For fixed k and n large, $n^{\underline{k}}$ behaves just like n^k. Specifically we have this.

Proposition 21. *Let k be a nonnegative integer. For n large we have $n^{\underline{k}} \sim n^k$.* $\qquad\qquad$ □

The proof is left for you (Exercise 9 on page 31).

4.5. Binomial coefficients. For fixed k, we have

$$\binom{n}{k} = \frac{n^{\underline{k}}}{k!} \sim \frac{n^k}{k!}$$

but when n, k, and $n - k$ all tend to infinity, we may apply Stirling's formula (Theorem 19):

$$\binom{n}{k} = \frac{n!}{k!(n-k)!} \sim \frac{\sqrt{2\pi n}(n/e)^n}{\sqrt{2\pi k}(k/e)^k \cdot \sqrt{2\pi(n-k)}((n-k)/e)^{n-k}} \tag{7}$$

$$= \sqrt{\frac{n}{2\pi k(n-k)}} \left(\frac{n^n}{k^k(n-k)^{n-k}} \right). \tag{8}$$

For example, if $k = \frac{1}{2}n$, then we can show that $\binom{n}{k} \asymp 2^n/\sqrt{n}$; see Exercise 39 on page 34.

On the other hand, consider a k satisfying $1 \ll k \ll \sqrt{n}$. The square-root term in (8) is asymptotic to $1/\sqrt{2\pi k}$ and the other term can be rewritten like this:

$$\left(\frac{n^n}{k^k(n-k)^{n-k}} \right) = \left(\frac{n}{k} \right)^k \left(\frac{n}{n-k} \right)^{n-k}$$

in which

$$\left(\frac{n}{n-k} \right)^{n-k} = \left(1 - \frac{k}{n} \right)^{-(n-k)} \sim e^k.$$

Combining all these we have, for $1 \ll k \ll \sqrt{n}$,

$$\binom{n}{k} \sim \frac{1}{\sqrt{2\pi k}} \left(\frac{ne}{k} \right)^k.$$

This implies (for n, k large enough with $1 \ll k \ll \sqrt{n}$) that $(\frac{n}{k})^k \le \binom{n}{k} \le (\frac{ne}{k})^k$. Indeed, this holds more broadly (and we prove this without the use of asymptotics).

Proposition 22. *For all $n, k \in \mathbb{Z}^+$ with $n \geq k$ we have*

$$\left(\frac{n}{k}\right)^k \leq \binom{n}{k} \leq \left(\frac{ne}{k}\right)^k.$$

Proof. Note that for any integer a with $0 \leq a < k$ we have

$$\frac{n-a}{k-a} \geq \frac{n}{k}.$$

Thus

$$\binom{n}{k} = \left(\frac{n}{k}\right) \cdot \left(\frac{n-1}{k-1}\right) \cdot \left(\frac{n-2}{k-2}\right) \cdots \left(\frac{n-k+1}{1}\right)$$

$$\geq \underbrace{\left(\frac{n}{k}\right) \cdot \left(\frac{n}{k}\right) \cdot \left(\frac{n}{k}\right) \cdots \left(\frac{n}{k}\right)}_{k \text{ terms}} = \left(\frac{n}{k}\right)^k.$$

For the upper bound, we use the fact that $k! \geq (k/e)^k$ (see Exercise 43 on page 35) to calculate:

$$\binom{n}{k} \leq \frac{n^k}{k!} \leq \frac{n^k}{(k/e)^k} = \left(\frac{ne}{k}\right)^k. \qquad \square$$

5. Hierarchy

When working with nasty algebraic expressions, it is important to get a feel for what is important and what is negligible. For example, consider this bit of ugliness (in which n is large):

$$n^3 \sqrt{\frac{\log \log n}{\log n}} + n^2 \log n. \tag{9}$$

Which term dominates: the first or the second? On the one hand, the first term has an n^3 and that's considerably bigger than n^2. On the other hand, the n^3 is multiplied by a term that's going to 0 while the n^2 is multiplied by $\log n$ which tends to infinity. Perhaps these extra factors offset the advantage n^3 has over n^2?

With time one can develop a good asymptotic "nose" and be able to "smell out" the important parts of complicated expressions. This takes some practice, but a good place to begin is by understanding the relative effects of the following sorts of terms all of which tend to infinity as $n \to \infty$.

- **Logarithmic**. These are expressions of the form $\log n$ or powers of logarithms (called *polylogs*) such as $(\log n)^3$ or $\sqrt{\log n}$.
- **Powers**. These are expressions of the form n^a for a positive exponent a. For example, a polynomial such as $n^3 - 2n^2 + 5$ is the sum of powers. Likewise, $\sqrt{2n}$ is a power term (the exponent is $\frac{1}{2}$).
- **Exponentials**. These are terms of the form b^n where $b > 1$. Examples include 2^n and $\exp n$.
- **Factorials**. Terms such as $n!$ or $n!!$. (See Exercise 23 on page 33 for the definition of *double factorial* $n!!$.)

To help you sort these out, we offer the following guiding result.

Proposition 23. *Let p, q, a be positive numbers with $a > 1$. For n large we have this:*

$$(\log n)^p \ll n^q \ll a^n \ll n! \ll n^n. \quad \square$$

The proof is not too difficult. We simply take the ratio of successive terms and see what happens as $n \to \infty$.

For example, to show that $(\log n)^p \ll n^q$ we take the limit

$$\lim_{n \to \infty} \frac{(\log n)^p}{n^q}.$$

Although this is an $\frac{\infty}{\infty}$ type of limit, it's difficult to apply Hôpital's rule (try it, you'll see why). Instead, to show that an expression tends to 0 we can show that its logarithm tends to $-\infty$. Taking logs we have

$$\log \left[\frac{(\log n)^p}{n^q} \right] = p \log \log n - q \log n.$$

Does *that* tend to $-\infty$? It does if we can show that $\log \log n \ll \log n$ and now Hôpital's rule helps:

$$\lim_{n \to \infty} \frac{\log \log n}{\log n} = \lim_{n \to \infty} \frac{1/(n \log n)}{1/n} = \lim_{n \to \infty} \frac{1}{\log n} = 0.$$

For the other $\ll$ inequalities, you can try the same method (and avail yourself of Stirling's formula when dealing with $n!$).

As you gain experience, you'll be able to discern (for example) that the first term in (9) dominates the second because the various logarithm terms are negligible compared to the extra factor of n in the first term.

6. Exercises

(1) TRUE or FALSE! For each of the following, determine if the statement is true or false.
   - (a) For large x, $\sin x$ is $O(1)$.
   - (b) For n large, n^2 is $O(n^3)$.
   - (c) For n large, n^2 is $\Theta(n^3)$.
   - (d) For n large, n^2 is $o(n^3)$.
   - (e) For fixed $k \in \mathbb{Z}$ with $k \geq 2$ and n large, n^k is $o(k^n)$.
   - (f) For fixed $a \in \mathbb{R}$ with $a > 0$ and n large, $\log n$ is $o(n^a)$.
   - (g) $1 \ll 10^{100}$.

(2) Suppose that $f(n) \sim g(n)$. Show that $f(n) - g(n) = o[g(n)]$.

(3) Show that $n^{1/n} \sim 1$ for n large.

(4) Give examples of functions f and g with $f \asymp g$ but $f \not\sim g$.

(5) Let f and g be nonzero polynomials. Show (for x large) that $f(x) \asymp g(x)$ if and only if f and g have the same degree.
   When does $f \sim g$?

(6) If a, b, c are numbers with $a < b$, then $a + c < b + c$. Asymptotically this isn't true. In particular, show that the following statement is false: If f, g, h are functions with $f \ll g$, then $f + h \ll g + h$.

(7) Let f be a polynomial. Show that $f(x) = o(e^x)$ for x large.

(8) A *rational function* is the ratio of two polynomials. That is, $f(x) = p(x)/q(x)$ where p and q are polynomials. Show that if f is a rational function then there is a real number c and an integer k such that $f(x) \sim cx^k$ (for x large). Assume that the numerator and denominator are not identically zero.

(9) Prove Proposition 21: For fixed k we have $n^{\underline{k}} \sim n^k$.

(10) For x small, show that

$$\frac{\log(1-x)}{x} \sim \frac{1}{x-1}.$$

(11) Let f be a function. Why is it possible to have $f \sim 2$, but not possible for $f \sim 0$?
   Then: Try to extend the definition of asymptotic equality to provide a meaning for $f \sim 0$.

(12) Without using Proposition 23, show, for n large, that $n^{100} \ll 1.01^n$.

31

(13) If f is a positive-valued function, then clearly $f(x) \to 0$ if and only if $\log f(x) \to -\infty$ (where $x \to \infty$). Use this to show that $x^{100} = o(e^x)$.

(14) For n large, consider the expressions $(\log n)^n$ and $n^{\log n}$. Which of the following relations (if any) hold between these two: $\ll$, $\asymp$, or $\gg$?

(15) Someone wrote $f(n) \sim n/\log n$ (for n large). Does this depend on the base of the logarithm?

Likewise, someone else wrote $g(n) \sim \log n / \log\log n$. Does this depend on the base of the logarithm?

(Note: An unsubscripted log is usually a base-e log unless otherwise noted.)

(16) A *factorion* is a positive integer that is equal to the sum of the factorials of its decimal digits. For example, 145 is a factorion because $1! + 4! + 5! = 145$.

Show that there are only finitely many factorions.

Hint: Let $f(n)$ be the sum of the factorials of the n's digits and prove that $f(n) = O(\log n)$.

Extra credit: Find all factorions.

(17) Find numbers a and b for which $1 - \cos x \sim ax^b$ (for x small).

(18) For a given $k \in \mathbb{N}$ let

$$f_k(n) = 1^k + 2^k + 3^k + \cdots + n^k.$$

Show that $f_k(n)$ is $\Theta(n^{k+1})$.

(19) Show that

$$\sum_{k=0}^{n} \frac{1}{\binom{n}{k}} = 2 + O(1/n).$$

(20) Suppose $f(x) = x^2 + O(x)$ and $g(x) = x^2 + O(x^{3/2})$. What can you say about $f(x) + g(x)$ when x is large ($x \to \infty$)? ... when x is small ($x \to 0$)?

(21) Show that

$$\sum_{k=1}^{n} \sqrt{k} = \Theta\left(n^{3/2}\right).$$

(22) Let f be an analytic function and let a be a real number. We know that for x near a that the ratio

$$\frac{f(x) - f(a)}{x - a}$$

is approximated by $f'(a)$. Rearranging, we can write this as

$$f(x) \approx f(a) + f'(a)(x - a).$$

How good is this approximation? That is, write

$$f(x) = f(a) + f'(a)(x - a) + ?$$

replacing the question mark with the best big-oh term you can.

(23) For positive, odd n, define the *double factorial* $n!!$ by

$$n!! = n \times (n - 2) \times (n - 4) \times \cdots \times 5 \times 3 \times 1.$$

More generally, we define $(-1)!! = 0!! = 1$ and then for any $n \in \mathbb{Z}^+$ put $n!! = n \cdot (n - 2)!!$.

Derive asymptotic formulas for $(2n)!!$ and $(2n - 1)!!$ where $n \in \mathbb{Z}^+$.

(24) Derive equation (6) (the sum of a geometric series) from the Generalized Binomial Theorem (Theorem 15).

(25) Prove that $\log(1 - x) \le -x$ for all $x < 1$.

(26) Prove Proposition 18: Let $n \in \mathbb{Z}^+$ and $x \in \mathbb{R}$ with $1 + \frac{x}{n} > 0$. Show that $\left(1 + \frac{x}{n}\right)^n \le e^x$.

(27) Derive the power series for $\sqrt{1 + x}$.

(28) Show that the hypothesis $f(n) = o(\sqrt{n})$ in Proposition 17 cannot be relaxed by showing that $\left(1 + \frac{\sqrt{n}}{n}\right)^n$ is not asymptotic to $\exp\left(\sqrt{n}\right)$. To do this, evaluate

$$\lim_{n \to \infty} \frac{\left(1 + \frac{1}{\sqrt{n}}\right)^n}{\exp\left(\sqrt{n}\right)}$$

and show that it does not equal 1.

(29) Show that $f = o(g)$ is not logically equivalent to $g = \omega(f)$. What condition might we place on the functions f and g to make this equivalence hold?

(30) Prove that $\sim$, is-asymptotically-equal-to, is an equivalence relation on functions.

(31) Find a pair of functions $f, g : \mathbb{N} \to \mathbb{R}$ for which both $f(n) \to \infty$ and $g(n) \to \infty$ as $n \to \infty$ and such that *none* of the following hold: $f \ll g$, $f \asymp g$, and $f \gg g$.

(32) In Proposition 9 we require the four functions to be positive-valued. Show that the conclusion is false if we drop this hypothesis.

(33) Let k be a positive integer and define

$$f_k(n) = \prod_{j=0}^{k} \frac{n-j}{n}.$$

Show that $f_k(n) \sim 1$.

Next, define

$$g(n) = \prod_{j=0}^{\lceil n/\log n \rceil} \frac{n-j}{n}.$$

Show that $g(n) \not\sim 1$ even though every term in the product is asymptotic to 1.

(34) Find functions f and g with $f \sim g$ but $\log f \not\sim \log g$.

(35) Find functions f and g such that $f \sim g$ but $e^f \not\sim e^g$.

(36) Verify equation (1): $\sqrt{n-1} = \sqrt{n} + o(1)$.

(37) Show that $\sqrt{n+1} - \sqrt{n} \sim 1/(2\sqrt{n})$.

Hint: Use the truncated Generalized Binomial Theorem (Proposition 16 on page 23) applied to $(1 + \frac{1}{n})^{1/2}$.

(38) For $n \in \mathbb{Z}^+$ the *harmonic number* H_n is

$$H_n = 1 + \frac{1}{2} + \frac{1}{3} + \cdots + \frac{1}{n}.$$

Prove that $H_n \sim \log n$.

Comment: A sharper version of this result is that

$$H_n = \log n + \gamma + o(1)$$

where γ is the *Euler-Mascheroni constant* whose approximate value is 0.5772.

(39) Show that $\binom{2n}{n} \sim 4^n / \sqrt{\pi n}$.

(40) For $x > 1$, define a function $\lambda(x)$ by the equation $x = \lambda(x) \log \lambda(x)$. Show that for x large

$$\lambda(x) \sim \frac{x}{\log x}.$$

Hint: Prove separately that $\lambda(x) > x/\log x$ and, given any $\varepsilon > 0$, then if x is sufficiently large we have $\lambda(x) < (1 + \varepsilon)x/\log x$.

(41) *(Exercise 40, continued.)* Substitute $x/\log x$ for the second $\lambda(x)$ in the equation $x = \lambda(x) \log \lambda(x)$ to find another approximation for the $\lambda(x)$ function and show, numerically, that this approximation is better than $\lambda(x) \approx x/\log x$.

(42) Use your answer to Exercise 41 to asymptotically solve for x in the equation $x^x = y$. Assume y is large.

(43) Let n be a positive integer. Stirling's formula asserts that $n! \sim \sqrt{2\pi n}(n/e)^n$. Without using this fact, prove by induction that $n! \geq (n/e)^n$.

(44) Let c be a positive real number and let $k \in \mathbb{N}$. Show that

$$\lim_{n \to \infty} \binom{n}{k} \left(\frac{c}{n}\right)^k \left(1 - \frac{c}{n}\right)^{n-k} = \frac{c^k}{k!} e^{-c}.$$

(45) Suppose that $\mathbf{f}, \mathbf{g} : \mathbb{N} \to \mathbb{R}^k$ be vector-valued functions. We wish to define $\mathbf{f} \sim \mathbf{g}$ (for n large).

One possibility is to write

$$\mathbf{f}(n) = \begin{bmatrix} f_1(n) \\ f_2(n) \\ \vdots \\ f_k(n) \end{bmatrix} \quad \text{and} \quad \mathbf{g}(n) = \begin{bmatrix} g_1(n) \\ g_2(n) \\ \vdots \\ g_k(n) \end{bmatrix}$$

and require that $f_j \sim g_j$ for all $j = 1, 2, \ldots, k$.

Give an example to show that this is a bad definition and propose a good definition.

(46) Suppose that f and g are differentiable functions with $f(x) \sim g(x)$ for x large. It is not necessarily the case that their derivatives are asymptotic. Find a pair of functions f and g so that $f \sim g$ but $f' \nsim g'$.

(47) Let f and g be integrable functions with $f(x) \sim g(x)$ for x large. Let

$$F(x) = \int_0^x f(t)\, dt \quad \text{and} \quad G(x) = \int_0^x g(t)\, dt.$$

It is not necessarily the case that if $f \sim g$ then their integrals are also asymptotic. Give examples of functions f and g so that $f \sim g$ but $F \nsim G$.

(48) Let f and g be integrable functions with $f(x) \sim g(x)$ for x large. Let

$$F(x) = \int_x^\infty f(t)\, dt \quad \text{and} \quad G(x) = \int_x^\infty g(t)\, dt.$$

We assume that $F(0)$ and $G(0)$ are both finite, so these are well-defined functions.

Prove that $F(x) \sim G(x)$.

(49) Let $f(x)$ be a polynomial. Show that
$$\sum_{k=1}^{n} f(k) \sim \int_{1}^{n} f(x)\,dx.$$
Then give an example of a nonnegative, increasing function for which this conclusion is false.

7. ANSWERS

(1) (a) True. (b) True. (c) False. (d) True. (e) True. (f) True. (g) False.

(2) Suppose $f(n) \sim g(n)$. Then $\lim f(n)/g(n) = 1$. Therefore

$$\lim \frac{f(n) - g(n)}{g(n)} = \lim \left(\frac{f(n)}{g(n)} - 1 \right) = 1 - 1 = 0$$

and therefore $f(n) - g(n) = o[g(n)]$.

(3) Since $\log\left(n^{1/n}\right) = \frac{1}{n}\log n$, we have

$$n^{1/n} = \exp\left[\frac{\log n}{n}\right] \to \exp(0) = 1$$

as $n \to \infty$. Therefore $n^{1/n} \sim 1$.

(4) Let $f(x) = x^2$ and $g(x) = 3x^2$. We have $f \asymp g$ but $f \nsim g$.

(5) ($\Rightarrow$) Suppose that f and g are polynomials of the same degree. Thus

$$f(x) = a_0 x^d + a_1 x^{d-1} + \cdots + a_d \quad \text{and}$$
$$g(x) = b_0 x^d + b_1 x^{d-1} + \cdots + b_d$$

where a_0 and b_0 are nonzero. Therefore

$$\lim_{x \to \infty} \left| \frac{f(x)}{g(x)} \right| = \left| \frac{a_0}{b_0} \right|.$$

Let $C = |a_0/b_0|$. Then for some $\varepsilon > 0$ and all x sufficiently large

$$C - \varepsilon \le |f(x)/g(x)| \le C + \varepsilon$$

and so $f \asymp g$.

($\Leftarrow$) If f and g have different degrees, then $\lim |f(x)/g(x)|$ is either 0 or ∞ and so we cannot bound their ratio. Therefore $f \nasymp g$. $\qquad\qquad\square$

We have $f \sim g$ exactly when f and g have the same degree and the same leading coefficient.

(6) Let $f(x) = x$, $g(x) = x^2$, and $h(x) = x^3$. Clearly $x \ll x^2$ but $x + x^3 \sim x^2 + x^3$.

(7) The proof is by induction on the degree of the polynomial. If f is a constant, then clearly $f(x)/e^x \to 0$ because $e^x \to \infty$ as $x \to \infty$.

37

Now assume that we have proved the result for polynomials of degree d. Let f be a polynomial of degree $d + 1$. Apply l'Hôpital's Rule:

$$\lim_{x \to \infty} \frac{f(x)}{e^x} = \lim_{x \to \infty} \frac{f'(x)}{e^x} = 0$$

because $f'(x)$ is a polynomial of degree d and so, by induction, $f'(x) = o(e^x)$. Therefore $f(x) = o(e^x)$. □

(8) Let

$$f(x) = \frac{p(x)}{q(x)} = \frac{a_s x^s + a_{s-1} x^{s-1} + \cdots + a_1 x + a_0}{b_t x^t + b_{t-1} x^{t-1} + \cdots + b_1 x + b_0}$$

where a_s and b_t are the leading terms (both not zero).

The numerator $p(x)$ is asymptotic to $a_s x^s$ and the denominator $q(x)$ is asymptotic to $b_t x^t$. By Proposition 10 we have

$$f(x) \sim \frac{a_s x^s}{b_t x^t} = c x^k$$

where $c = a_s / b_t$ and $k = s - t$.

(9) We prove that $n^{\underline{k}} \sim n^k$ by induction on k. The basis case $k = 0$ is trivial (both terms are identically equal to 1). Suppose the result has been proved for a given value of k and we work to show that $n^{\underline{k+1}} \sim n^{k+1}$.

By induction, $n^{\underline{k}} \sim n^k$. Multiply both sides by $n - k$ and we have

$$n^{\underline{k+1}} = n^{\underline{k}}(n - k) \sim n^k(n - k) \sim n^k \cdot n = n^{k+1}$$

as required.

(10) To show $\log(1 - x)/x \sim 1/(x - 1)$ we divide and take the limit as $x \to 0$:

$$\lim_{x \to 0} \frac{\log(1 - x)}{x} \div \frac{1}{x - 1} = \lim_{x \to 0} \frac{(x - 1)\log(1 - x)}{x}$$

$$= \lim_{x \to 0} \frac{[(x - 1)\log(1 - x)]'}{x'}$$

$$= \lim_{x \to 0}(x - 1)\left(\frac{-1}{1 - x}\right) + \log(1 - x)$$

$$= 1.$$

(11) When we write $f \sim 2$ it simply means that the limit of $f(x)/2$ is 1. For example, if $f(x) = (2x - 1)/(x + 1)$, then we have $f \sim 2$.

38

However, for $f \sim 0$ we either have $f(x)/0$ or $0/f(x)$ tending to 1, but that's clearly impossible.

If there were a need, one might extend the definition of asymptotic equality and declare $f \sim 0$ to mean that $f(x)$ exactly equals 0 for all x sufficiently large.

(12) Both n^{100} and 1.01^n tend to infinity for large n. To show that their ratio tends to zero it is enough to show that

$$\log \left[\frac{n^{100}}{1.01^n} \right] \to -\infty.$$

This evaluates to

$$100 \log n - n \log 1.01.$$

Both of these terms also tend to ∞ but to show that the second term dominates we evaluate the limit via Hôpital's rule:

$$\lim_{n \to \infty} \frac{100 \log n}{n \log 1.01} = \lim_{n \to \infty} \frac{100/n}{\log 1.01} = 0.$$

Therefore $100 \log n - n \log 1.01 \to -\infty$ and the result follows.

(13) We want to show that $x^{100}/e^x \to 0$. Taking logs we have

$$\log \left(\frac{x^{100}}{e^x} \right) = 100 \log x - x \to -\infty.$$

(14) To understand the behavior of the ratio

$$\frac{(\log n)^n}{n^{\log n}}$$

we take the logarithm of numerator and denominator:

$$\log \left[(\log n)^n \right] = n \log \log n \quad \text{and}$$
$$\log \left[n^{\log n} \right] = (\log n)^2.$$

By Proposition 23, $n^1 \gg (\log n)^2$ and so

$$n \log \log n - (\log n)^2 \to +\infty$$

and we conclude that $(\log n)^n \gg n^{\log n}$.

As a quick computational check, for $n = 100$ we have $(\log n)^n \approx 2.1 \times 10^{66}$ and $n^{\log n} \approx 1.6 \times 10^9$.

(15) The statement $f(n) \sim n/\log n$ depends on the base of the logarithm. For distinct positive numbers a and b let

$$f_a(n) = \frac{n}{\log_a n} \quad \text{and} \quad f_b(n) = \frac{n}{\log_b n}$$

If the base of the log didn't matter, we'd have $f_a \sim f_b$. However

$$\frac{f_a(n)}{f_b(n)} = \frac{n/\log_a n}{n/\log_b n} = \frac{\log_b n}{\log_a n} = \log_b a \neq 1.$$

On the other hand, $\log n / \log \log n$ is asymptotically independent of the base of the logs. Observe:

$$\frac{\log_a n}{\log_a \log_a n} = \frac{\log n / \log a}{(\log \log_a n)/\log a}$$

$$= \frac{\log n}{\log(\log n / \log a)}$$

$$= \frac{\log n}{\log \log n - \log \log a}$$

$$\sim \frac{\log n}{\log \log n}.$$

(16) For a positive integer n let $f(n)$ be the sum of the factorials of n's digits. Since n has $\lceil \log_{10} n \rceil$ digits, we have $f(n) \leq 9! \lceil \log_{10} n \rceil = O(\log n)$. A factorion satisfies $f(n) = n$ and since $n \gg \log n$, the equation $f(n) = n$ is false for all n sufficiently large, i.e., for all but finitely many n.

There are only four factorions (as can be verified by a computer program). They are $1, 2, 145$, and 40585.

(17) Since $\cos x = 1 - \frac{x^2}{2!} + O(x^4)$ we have $1 - \cos x \sim \frac{1}{2}x^2$, so $a = \frac{1}{2}$ and $b = 2$.

(18) On the one hand

$$f_k(n) = 1^k + 2^k + 3^k + \cdots + n^k$$

$$\leq n^k + n^k + n^k + \cdots + n^k = n^{k+1}$$

so $f_k(n)$ is $O(n^{k+1})$.

40

On the other hand
$$f_k(n) = 1^k + 2^k + 3^k + \cdots + n^k$$
$$\geq \underbrace{\left(\frac{n}{2}\right)^k + \cdots + \left(\frac{n}{2}\right)^k}_{n/2 \text{ terms}}$$
$$= \frac{n}{2} \cdot \frac{n^k}{2^k} = \frac{n^{k+1}}{2^k}$$

and so $f_k(n)$ is $\Omega(n^{k+1})$. Therefore $f_k(n)$ is $\Theta(n^{k+1})$.
See also Exercise 49.

(19) In the sum $\sum \binom{n}{k}^{-1}$, the $k = 0$ and $k = n$ terms give $1 + 1 = 2$. Both the $k = 1$ and $k = n - 1$ terms equal $1/n$. Thus

$$\sum_{k=0}^{n} \frac{1}{\binom{n}{k}} = 2 + \frac{2}{n} + \sum_{k=2}^{n-2} \frac{1}{\binom{n}{k}}.$$

The remaining $n - 4$ terms (from $k = 2$ to $k = n - 2$) are bounded by $1/\binom{n}{2}$ so their sum is bounded by $(n - 4)/\binom{n}{2} = O(1/n)$. The result follows.

(20) For x large, $f(x) + g(x) = 2x^2 + O(x^{3/2})$. For x small, $f(x) + g(x) = 2x^2 + O(x)$.

(21) We have the following simple upper bound.

$$\sum_{k=1}^{n} \sqrt{k} \leq \sum_{k=1}^{n} \sqrt{n} = n^{3/2}.$$

For the lower bound, just consider half of the terms:

$$\sum_{k=1}^{n} \sqrt{k} \geq \sum_{k=\lfloor n/2 \rfloor}^{n} \sqrt{k} \geq \sum_{k=\lfloor n/2 \rfloor}^{n} \sqrt{\lfloor n/2 \rfloor} \sim \frac{n^{3/2}}{2\sqrt{2}}.$$

(22) Apply Theorem 13 to the function $g(x) = f(x - a)$ to get

$$f(x) = f(a) + f'(a)(x - a) + O\left((x - a)^2\right).$$

(23) For even integers:

$$(2n)!! = [2n] \cdot [2(n-1)] \cdot [2(n-2)] \cdots [2(1)] = 2^n n!$$
$$\sim 2^n \sqrt{2\pi n}(n/e)^n = \sqrt{2\pi n} \left(\frac{2n}{e}\right)^n.$$

For odd integers:

$$(2n - 1)!! = \frac{(2n)!}{(2n)!!} \sim \frac{\sqrt{2\pi 2n}(2n/e)^{2n}}{\sqrt{2\pi n}(2n/e)^n} = \sqrt{2} \left(\frac{2n}{e}\right)^n.$$

41

(24) Substitute -1 for α and $-x$ for x in the Generalized Binomial Theorem to give

$$(1-x)^{-1} = \sum_{k=0}^{\infty} \binom{-1}{k} (-x)^k. \tag{10}$$

The binomial coefficient $\binom{-1}{k}$ is

$$\binom{-1}{k} = \frac{(-1)(-2)(-3)\cdots(-k)}{k!} = (-1)^k.$$

Substituting this into (10) yields

$$(1-x)^{-1} = 1 + x + x^2 + x^3 + \cdots.$$

(25) That $\log(1-x) \le -x$ for all $x < 1$ is easy to see from Figure 4. Here's an analytic proof.

Let $f(x) = \log(1-x) + x$. We need to show that $f(x) \le 0$ for all $x < 1$.

Note that $f(0) = 0$ and $f'(x) = 1 - (1-x)^{-1}$. If $x < 0$ then $f'(x) > 0$ so f is increasing on $(-\infty, 0]$; therefore $f(x) \le 0$ for $x \le 0$. For $0 \le x < 1$ note that $f'(x) < 0$ so f is decreasing on $[0,1)$ and therefore $f(x) \le 0$ on this interval as well.

(26) From Proposition 14 we know that $1 + x \le e^x$. Substituting x/n for x gives

$$1 + \frac{x}{n} \le e^{x/n}.$$

Since $1 + \frac{x}{n} > 0$ raising both sides to the n preserves the inequality:

$$\left(1 + \frac{x}{n}\right)^n \le \left(e^{x/n}\right)^n = e^x.$$

(27) We use the Generalized Binomial Theorem:

$$\sqrt{1+x} = (1+x)^{\frac{1}{2}} = \sum_{k=0}^{\infty} \binom{1/2}{k} x^k.$$

Note that $\binom{1/2}{0} = 1$ and for $k > 0$

$$\binom{1/2}{k} = \frac{(\frac{1}{2})(\frac{-1}{2})(\frac{-3}{2})\cdots(\frac{-2k+3}{2})}{k!} = (-1)^{k-1}\frac{(2k-3)!!}{2^k k!}$$

42

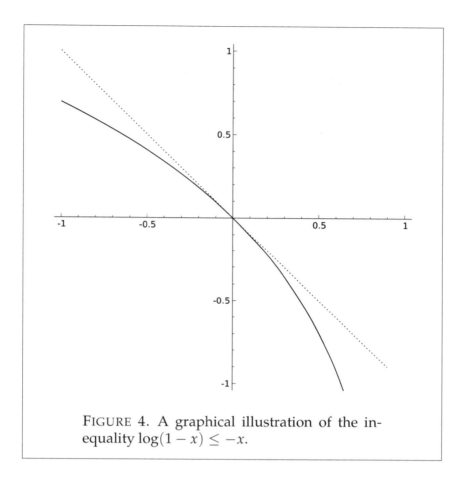

FIGURE 4. A graphical illustration of the inequality $\log(1 - x) \le -x$.

where the double factorial is defined in Exercise 23. Therefore

$$\sqrt{1+x} = 1 + \sum_{k=1}^{\infty} (-1)^{k-1} \frac{(2k-3)!!}{2^k k!} x^k$$

$$= 1 + \frac{1}{2}x - \frac{1}{8}x^2 + \frac{1}{16}x^3 - \frac{5}{128}x^4 + \frac{7}{256}x^5 - \cdots .$$

(28) Let $f(n) = \left(1 + \frac{1}{\sqrt{n}}\right)^n \Big/ \exp\left(\sqrt{n}\right)$. Then

$$\log f(n) = n \log\left(1 + \frac{1}{\sqrt{n}}\right) - \sqrt{n}. \tag{11}$$

43

We have

$$\log\left(1 + \frac{1}{\sqrt{n}}\right) = \frac{1}{\sqrt{n}} - \frac{1}{2n} + O(n^{-3/2}).$$

Substituting this in (11) gives

$$\log f(n) = \left(\sqrt{n} - \frac{1}{2} + O(n^{-1/2})\right) - \sqrt{n}$$

which gives $\lim f(n) = e^{-1/2}$.

Therefore $\left(1 + \frac{1}{\sqrt{n}}\right)^n$ is not asymptotic to $\exp\sqrt{n}$.

(29) Let $f(n) = -n$ and $g(n) = n^2 + 1$. Then $f = o(g)$ [because $-n/(n^2 + 1) \to 0$ as $n \to \infty$] but $g \neq w(f)$ because $g/f \not\to +\infty$. If we restrict the two functions to take on only positive values, then the equivalence holds.

(30) (a) $\sim$ is reflexive: $f \sim f$ because $\lim \frac{f(x)}{f(x)} = 1$.

(b) $\sim$ is symmetric: Given $f \sim g$ we have $\lim \frac{f(x)}{g(x)} = 1$.

This implies that $\lim \frac{g(x)}{f(x)} = 1$ (because $x \mapsto 1/x$ is continuous at $x = 1$) and therefore $g \sim f$.

(c) $\sim$ is transitive: Give $f \sim g$ and $g \sim h$, we have $\lim \frac{f(x)}{g(x)} = 1$ and $\lim \frac{g(x)}{h(x)} = 1$. Therefore

$$\lim \frac{f(x)}{h(x)} = \lim \frac{f(x)g(x)}{g(x)h(x)}$$

$$= \left[\lim \frac{f(x)}{g(x)}\right] \cdot \left[\lim \frac{g(x)}{h(x)}\right] = 1 \cdot 1 = 1$$

and so $f \sim h$. $\qquad\square$

(31) Let

$$f(n) = n^2 \qquad \text{and}$$

$$g(n) = n + [1 - (-1)^n]\, n^3 = \begin{cases} n & \text{if } n \text{ is even} \\ n + 2n^3 & \text{if } n \text{ is odd.} \end{cases}$$

Then $f(n)/g(n)$ grows arbitrarily large for n even and arbitrarily small for n odd. Hence none of $f \ll g$, $f \asymp g$, nor $f \gg g$ hold.

44

(32) Let

$$f_1(n) = \binom{n}{2}, \qquad\qquad f_2(n) = 1 - \binom{n}{2},$$

$$g_1(n) = \frac{n^2}{2}, \quad \text{and} \qquad g_2(n) = -\binom{n}{2}.$$

Note that $f_1 \sim g_1$ and $f_2 \sim g_2$. However

$$\lim_{n\to\infty} \frac{f_1(n) + f_2(n)}{g_1(n) + g_2(n)} = \lim_{n\to\infty} \frac{1}{n/2} = 0$$

and so $f_1 + f_2 \not\sim g_1 + g_2$.

(33) We show that $f_k(n) \sim 1$ by induction on k.

For $k = 1$, $f_1(n) = \frac{n}{n} \cdot \frac{n-1}{n} = \frac{n-1}{n} \sim 1$.

So assume the result have been proved for smaller values of k and note that

$$f_k(n) = f_{k-1}(n) \cdot \frac{n-k}{n}.$$

Since $\frac{n-k}{n} \sim 1$, it follows that $f_k(n) \sim f_{k-1}(n) \sim 1$ (using Proposition 12).

Now we show that $g(n) \not\sim 1$.

First, note that the smallest factor in $g(n)$ is

$$\frac{n - \lceil n/\log n \rceil}{n} = 1 - O\left(\frac{1}{\log n}\right) \sim 1$$

and thus all the factors are asymptotic to 1.

Let's break the product for $g(n)$ into two nearly equal size ranges. Let $r = \lceil \frac{1}{2}n/\log n \rceil$ and write

$$g(n) = \prod_{j=0}^{r-1} \frac{n-j}{n} \times \prod_{j=r}^{\lceil n/\log n\rceil} \frac{n-j}{n}$$

All the terms in the first product at at most 1 and all the terms in the second product are at most $(n - r)/n$. Therefore

$$g(n) \leq \left(1 - \frac{r}{n}\right)^{r + O(1)}.$$

(The $O(1)$ term in the exponent accounts for the fact that, perhaps, $\lceil n/\log n \rceil$ is odd and there are not exactly r terms in the second product.)

45

Taking logs we have

$$\log g(n) \leq (r + O(1)) \log \left(1 - \frac{r}{n}\right)$$

$$= (r + O(1)) \left(-\frac{r}{n} + O(r^2/n^2)\right)$$

$$= \left(\frac{n}{2 \log n} + O(1)\right) \left(-\frac{\frac{1}{2} n / \log n}{n} + O(1/\log^2 n)\right)$$

$$= -\frac{n}{4 \log^2 n} + O\left(\frac{1}{\log n}\right) + O\left(\frac{n}{\log^3 n}\right) + o(1)$$

$$\sim -\frac{n}{4 \log^2 n} \to -\infty.$$

Therefore $g(n) \to 0$ as $n \to \infty$ and thus $g(n) \not\sim 1$.

(34) Let $f(n) = 1 + \frac{1}{n}$ and $g(n) = 1 + \frac{1}{n^2}$. Clearly $f(n) \sim g(n)$ for n large. However

$$\log f(n) = \log \left(1 + \frac{1}{n}\right) = \frac{1}{n} + O(n^{-2}) \quad \text{and}$$

$$\log g(n) = \log \left(1 + \frac{1}{n^2}\right) = \frac{1}{n^2} + O(n^{-4})$$

and so $\log f(n) \not\sim \log g(n)$.

(35) Let $f(n) = 1 + \log n$ and $g(n) = \log n$. Note that

$$\lim_{n \to \infty} \frac{f(n)}{g(n)} = \lim_{n \to \infty} \frac{1 + \log n}{\log n} = \lim_{n \to \infty} \left(\frac{1}{\log n} + 1\right) = 1$$

and so $f \sim g$. However

$$e^{f(n)} = e^{\log n} = n \quad \text{and} \quad e^{g(n)} = e^{1 + \log n} = en$$

and so $f \not\sim g$.

(36) We must show that $\sqrt{n} - \sqrt{n-1} \to 0$ as $n \to \infty$. Notice that

$$\left(\sqrt{n} - \sqrt{n-1}\right) \left(\sqrt{n} + \sqrt{n-1}\right) = n - (n-1) = 1.$$

Therefore

$$\lim_{n \to \infty} \left(\sqrt{n} - \sqrt{n-1}\right) = \lim_{n \to \infty} \frac{1}{\sqrt{n} + \sqrt{n-1}} = 0.$$

(37) Take a factor of $\sqrt{n}$ out of the left side:

$$\sqrt{n+1} - \sqrt{n} = \sqrt{n} \left[(1 + \tfrac{1}{n})^{1/2} - 1\right]. \tag{12}$$

Apply Theorem 15 to give

$$\left(1+\tfrac{1}{n}\right)^{1/2} = 1 + \binom{\frac{1}{2}}{1}\frac{1}{n} + \binom{\frac{1}{2}}{2}\frac{1}{n^2} + \cdots$$

$$= 1 + \frac{1}{2n} + O(n^{-2}).$$

Substitute into (12):

$$\sqrt{n+1} - \sqrt{n} = \sqrt{n}\left[1 + \frac{1}{2n} + O(n^{-2}) - 1\right]$$

$$= \frac{1}{2\sqrt{n}} + O(n^{-3/2}) \sim \frac{1}{2\sqrt{n}}.$$

(38) We can approximate $H_n = 1 + \tfrac{1}{2} + \tfrac{1}{3} + \cdots + \tfrac{1}{n}$ by the integral $\int \frac{1}{x}\,dx$. Refer to Figure 5 on the following page. Using the upper portion of the figure, we have

$$H_n \geq \int_1^{n+1} \frac{1}{x}\,dx = \log(n+1)$$

Next, using the lower diagram we have

$$\frac{1}{2} + \frac{1}{3} + \cdots + \frac{1}{n} \leq \int_1^n \frac{1}{x}\,dx = \log n$$

and so $H_n \leq \log n + 1$. In summary:

$$\log(n+1) \leq H_n \leq 1 + \log n.$$

Divide this by $\log n$

$$\frac{\log(n+1)}{\log n} \leq \frac{H_n}{\log n} \leq \frac{1 + \log n}{\log n}$$

and take limits as $n \to \infty$. Since both $\log(n+1)/\log n$ and $(1 + \log n)/\log n$ tend to 1, we have

$$\lim_{n\to\infty} \frac{H_n}{\log n} = 1$$

and so $H_n \sim \log n$.

(39) Using Stirling's formula we have $n! \sim \sqrt{2\pi n}(n/e)^n$ and $(2n)! \sim \sqrt{4\pi n}(2n/e)^{2n}$. We calculate

$$\binom{2n}{n} = \frac{(2n)!}{n!n!} \sim \frac{\sqrt{4\pi n}(2n/e)^{2n}}{2\pi n(n/e)^{2n}} = \frac{2^{2n}}{\sqrt{\pi n}}.$$

47

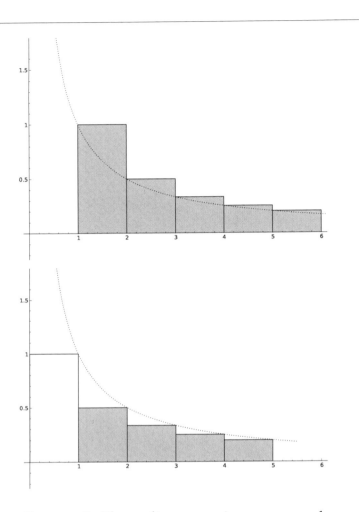

FIGURE 5. These diagrams give upper and lower bounds for $H_5 = 1 + \frac{1}{2} + \frac{1}{3} + \frac{1}{4} + \frac{1}{5}$. The dotted curves in these figures are the graph of $y = 1/x$. The top figure illustrates that $H_5 \geq \int_1^6 \frac{1}{x}\,dx$. The bottom figure shows that $H_5 \leq 1 + \int_1^5 \frac{1}{x}\,dx$.

(40) For easier reading, we write λ in place of $\lambda(x)$.

First we show that $\lambda > x/\log x$. Suppose, for contradiction, that $\lambda \leq x/\log x$. Then

$$x = \lambda \log \lambda \leq \frac{x}{\log x} \log \left(\frac{x}{\log x} \right)$$

$$= \frac{x}{\log x} (\log x - \log \log x)$$

$$= x - \frac{\log \log x}{\log x} < x. \qquad \Rightarrow \Leftarrow$$

Second we show that given $\varepsilon > 0$, for x sufficiently large, we have $\lambda < (1+\varepsilon)x/\log x$. Again, for sake of contradiction, suppose $\lambda \geq (1+\varepsilon)x/\log x$. Then

$$x = \lambda \log \lambda \geq (1+\varepsilon)\frac{x}{\log x} \log \left[(1+\varepsilon)\frac{x}{\log x} \right]$$

$$= (1+\varepsilon)\frac{x}{\log x} \left[\log x + \log(1+\varepsilon) - \log \log x \right]$$

$$= (1+\varepsilon)x + x \left[\frac{\log(1+\varepsilon) - \log \log x}{\log x} \right]$$

$$= (1+\varepsilon)x + o(x) \geq (1+\varepsilon/2)x. \qquad \Rightarrow \Leftarrow$$

Thus, for any $\varepsilon > 0$ we have, for x sufficiently large,

$$\frac{x}{\log x} < \lambda(x) < (1+\varepsilon)\frac{x}{\log x}$$

and therefore

$$\lim_{x \to \infty} \frac{\lambda(x)}{x/\log x} = 1$$

giving $\lambda(x) \sim x/\log x$.

(41) Into $x = \lambda(x) \log \lambda(x)$ we substitute $\lambda(x) \sim x/\log x$ to get

$$x \approx \lambda(x) \log \left(\frac{x}{\log x} \right) = \lambda(x)(\log x - \log \log x)$$

from which we derive

$$\lambda(x) \approx \frac{x}{\log x - \log \log x}.$$

49

If we take $x = 10^{10}$ we have the following results:

$$\lambda(x) = 4.9928 \times 10^8$$
$$x/\log x = 4.3429 \times 10^8$$
$$x/(\log x - \log\log x) = 5.0278 \times 10^8$$

The $x/\log x$ approximation is about 13% too small whereas the $x/(\log x - \log\log x)$ approximation roughly 0.7% too large.

Note that the two approximations are asymptotically equal to each other (and to $\lambda(x)$).

(42) Note that solving $y = x^x$ is equivalent to solving $\log y = x \log x$. Using the asymptotic formula from Exercise 41 we have

$$x = \frac{\log y}{\log\log y - \log\log\log y}.$$

When $y = 100^{100}$ (so the solution to $y = x^x$ is simply $x = 100$) the formula gives the value 106.63.

When $y = g^g$ where $g = 10^{100}$ (a googol) the formula gives a value that is off by about 0.01%.

(43) We prove $n! \geq (n/e)^n$ by induction on n. The case $n = 1$ is trivial, so we assume the result has been shown for $n = k$. We have

$$(k+1)! = (k+1)k! \geq (k+1)(k/e)^k = \frac{(k+1)(k+1)^k k^k}{(k+1)^k e^k}$$

$$= \left(\frac{k}{k+1}\right)^k \left(\frac{(k+1)^{k+1}}{e^k}\right) \geq \frac{1}{e}\left(\frac{(k+1)^{k+1}}{e^k}\right)$$

$$= \left(\frac{k+1}{e}\right)^{k+1}$$

where we used Proposition 18 to justify $\left(\frac{k}{k+1}\right)^k \geq \frac{1}{e}$ because $\left(1 + \frac{1}{k}\right)^k \leq e^1$.

(44) For $c > 0$ and fixed $k \in \mathbb{N}$ note that

$$\binom{n}{k}\left(\frac{c}{n}\right)^k \sim \frac{n^k}{k!} \cdot \frac{c^k}{n^k} \sim \frac{c^k}{k!} \qquad \text{and}$$

$$\left(1 - \frac{c}{n}\right)^{n-k} \sim \left(1 - \frac{c}{n}\right)^n \sim e^{-c}.$$

Multiplying these gives the result.

(45) Let

$$\mathbf{f}(n) = \begin{bmatrix} n^2 \\ \log n \end{bmatrix} \quad \text{and} \quad \mathbf{g}(n) = \begin{bmatrix} n^2 + 1 \\ 1/n \end{bmatrix}.$$

These are functions we should consider asymptotically equal, but although $f_1 \sim g_1$, it is not the case that $f_2 \sim g_2$.

One idea is to require the vectors $\mathbf{f}(n)$ and $\mathbf{g}(n)$ to have asymptotically the same magnitude and direction. Thus we say $\mathbf{f} \sim \mathbf{g}$ provided:

$$\lim_{n \to \infty} \frac{\|\mathbf{f}(n)\|}{\|\mathbf{g}(n)\|} = \lim_{n \to \infty} \frac{\mathbf{f}(n)}{\|\mathbf{f}(n)\|} \cdot \frac{\mathbf{g}(n)}{\|\mathbf{g}(n)\|} = 1.$$

(46) Let $f(x) = 1 + 1/x$ and $g(x) = 1 + 1/x^2$. Clearly $f \sim g$. But $f'(x) = -1/x^2$ and $g'(x) = -2/x^3$ and these functions are not asymptotic.

(47) Let

$$f(x) = \frac{1}{(x+1)^2} + e^{-x} \quad \text{and} \quad g(x) = \frac{1}{(x+1)^2}.$$

Note that

$$\frac{f(x)}{g(x)} = \frac{(x+1)^{-2} + e^{-x}}{(x+1)^{-2}} = 1 + e^{-x}(1 + x^2) \to 1$$

as $x \to \infty$, and so $f \sim g$.
    However,

$$\lim_{x \to \infty} F(x) = \int_0^\infty f(t)\, dt = 2$$

$$\lim_{x \to \infty} G(x) = \int_0^\infty g(t)\, dt = 1$$

and so $F \not\sim G$.

(48) Observe that

$$F(x) = \int_x^\infty f(t)\, dt = \int_0^\infty f(t)\, dt - \int_0^x f(t)\, dt$$

$$= F(0) - \int_0^x f(t)\, dt.$$

From this it follows that

$$\lim_{x \to \infty} F(x) = F(0) - \lim_{x \to \infty} \int_0^x f(t)\, dt = F(0) - F(0) = 0$$

51

and that $F'(x) = -f(x)$ by the Fundamental Theorem of Calculus. Likewise we have $\lim_{x\to\infty} G(x) = 0$ and $G'(x) = -g(x)$.

Therefore, by l'Hôpital's Rule,

$$\lim_{x\to\infty} \frac{F(x)}{G(x)} = \lim_{x\to\infty} \frac{-f(x)}{-g(x)} = 1$$

and we conclude that $F \sim G$.

(49) First, note that it is enough to show the result for $f(x) = x^a$ for $a \in \mathbb{N}$. Our goal reduces to showing that

$$\sum_{k=1}^{n} k^a \sim \int_1^n x^a\, dx \sim \frac{n^{a+1}}{a+1}.$$

For example,

$$\sum_{k=1}^{n} k^4 = \tfrac{1}{5}n^5 + \tfrac{1}{2}n^4 + \tfrac{1}{3}n^3 - \tfrac{1}{30}n \sim \tfrac{1}{5}n^5.$$

Begin by examining Figure 6. The dotted curve is the graph of $y = x^a$ (illustrated with $a = 2$). From the upper portion of the figure we have

$$\sum_{k=1}^{n} k^a \le \int_1^{n+1} x^a\, dx = \frac{(n+1)^{a+1}}{a+1} - \frac{1}{a+1} \sim \frac{n^{a+1}}{a+1}.$$

From the lower portion of the figure we have

$$1 + \sum_{k=2}^{n+1} k^a \ge 1 + \int_1^{n+1} x^a\, dx$$

$$\Rightarrow \quad \sum_{k=1}^{n} k^a \ge 1 - (n+1)^a + \int_1^{n+1} x^a\, dx$$

$$= 1 - (n+1)^a + \frac{(n+1)^{a+1} - 1}{a+1} \sim \frac{n^{a+1}}{a+1}.$$

Therefore $\sum_{k=1}^{n} k^a \sim n^{a+1}/(a+1)$ and the result follows.

Finally, consider $f(x) = 2^x$. On the one hand,

$$\sum_{k=1}^{n} f(x) = \sum_{k=1}^{n} 2^k = 2^{n+1} - 2 \sim 2 \cdot 2^n$$

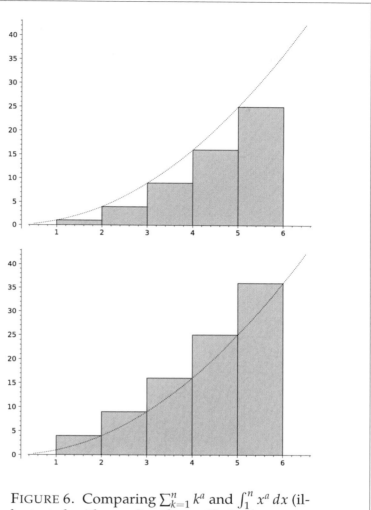

FIGURE 6. Comparing $\sum_{k=1}^{n} k^a$ and $\int_{1}^{n} x^a \, dx$ (illustrated with $a = 2$ and $n = 5$). In the upper portion the rectangles represent $\sum_{k=1}^{n} k^a$ and in the lower portion the rectangles represent $\sum_{k=2}^{n+1} k^a$.

but on the other hand
$$\int_1^n f(x)\,dx = \int_1^n 2^x\,dx = \frac{2^n - 2}{\log 2} \sim \frac{1}{\log 2} \cdot 2^n$$
and therefore these are not asymptotically equal.

References

[1] Ronald L. Graham, Donald E. Knuth, and Oren Patashnik. *Concrete Mathematics: A Foundation for Computer Science*. Addison-Wesley, second edition, 1994.

[2] Edward R. Scheinerman. *Mathematical Notation: A Guide for Engineers and Scientists*. Create Space, 2011.

[3] Edward R. Scheinerman. *Mathematics: A Discrete Introduction*. Brooks/Cole, third edition, 2012.

[4] Joel Spencer. *Asymptopia*. American Mathematical Society, 2014. With Laura Florescu.

Notation

$n!$: factorial, 24
$n^{\underline{k}} = (n)_k$: falling factorial, 26
$n!!$: double factorial, e.g., $7!! = 7 \times 5 \times 3 \times 1$, 33
$f \gg g$: asymptotically much greater than, 15
$f \ll g$: asymptotically much less than, 14
$f \asymp g$: same asymptotic growth, 13
$f \sim g$: asymptotic equality, 9
$\binom{\alpha}{k}$: (generalized) binomial coefficient, 22

exp: exponential function, e^x, 20

H_n: harmonic number $1 + \frac{1}{2} + \cdots + \frac{1}{n}$, 34

log: base-e logarithm, 20

$\mathbb{N}$: the natural numbers (including 0), 10

$O(f)$: big oh, 10
$o(f)$: little oh, 13
$\Omega(f)$: big omega, 12
$\omega(f)$: little omega, 14

$\pi(n)$: number of primes at most n, 7

$\mathbb{R}$: the real numbers, 9

$\Theta(f)$: big theta, 12

$\mathbb{Z}$: the integers, 9
$\mathbb{Z}^+$: the positive integers, 9

INDEX

Made in the USA
Columbia, SC
21 January 2021

31353412R00033